MIX
Papier aus verantwortungsvollen Quellen
Paper from responsible sources
FSC® C105338

Dr. Pathik Shah
Kalpana Pandey

Development of Packaging Film Using Microcrystalline Cellulose and Pro-Oxidative Additive Using Blown Film Technique

Anchor Academic Publishing

Shah, Pathik, Pandey, Kalpana: Development of Packaging Film Using Microcrystalline
Cellulose and Pro-Oxidative Additive Using Blown Film Technique, Hamburg, Anchor
Academic Publishing 2017

Buch-ISBN: 978-3-96067-188-6
PDF-eBook-ISBN: 978-3-96067-688-1
Druck/Herstellung: Anchor Academic Publishing, Hamburg, 2017

Bibliografische Information der Deutschen Nationalbibliothek:
Die Deutsche Nationalbibliothek verzeichnet diese Publikation in der Deutschen
Nationalbibliografie; detaillierte bibliografische Daten sind im Internet über
http://dnb.d-nb.de abrufbar.

Bibliographical Information of the German National Library:
The German National Library lists this publication in the German National Bibliography.
Detailed bibliographic data can be found at: http://dnb.d-nb.de

All rights reserved. This publication may not be reproduced, stored in a retrieval system
or transmitted, in any form or by any means, electronic, mechanical, photocopying,
recording or otherwise, without the prior permission of the publishers.

Das Werk einschließlich aller seiner Teile ist urheberrechtlich geschützt. Jede Verwertung
außerhalb der Grenzen des Urheberrechtsgesetzes ist ohne Zustimmung des Verlages
unzulässig und strafbar. Dies gilt insbesondere für Vervielfältigungen, Übersetzungen,
Mikroverfilmungen und die Einspeicherung und Bearbeitung in elektronischen Systemen.

Die Wiedergabe von Gebrauchsnamen, Handelsnamen, Warenbezeichnungen usw. in
diesem Werk berechtigt auch ohne besondere Kennzeichnung nicht zu der Annahme,
dass solche Namen im Sinne der Warenzeichen- und Markenschutz-Gesetzgebung als frei
zu betrachten wären und daher von jedermann benutzt werden dürften.

Die Informationen in diesem Werk wurden mit Sorgfalt erarbeitet. Dennoch können
Fehler nicht vollständig ausgeschlossen werden und die Diplomica Verlag GmbH, die
Autoren oder Übersetzer übernehmen keine juristische Verantwortung oder irgendeine
Haftung für evtl. verbliebene fehlerhafte Angaben und deren Folgen.

Alle Rechte vorbehalten

© Anchor Academic Publishing, Imprint der Diplomica Verlag GmbH
Hermannstal 119k, 22119 Hamburg
http://www.diplomica-verlag.de, Hamburg 2017
Printed in Germany

Development of Packaging Film Using Microcrystalline Cellulose and Pro-Oxidative Additive Using Blown Film Technique

Pathik Shah[*], Kalpana Pandey

Central Institute of Plastics Engineering and Technology (CIPET), Vatva GIDC, Ahmedabad, India

Email address: pathikas@gmail.com

ABSTRACT

The purpose of this study is to develop degradable Cellulose based packaging film with improved mechanical properties. A series of Linear Low Density Polyethylene (LLDPE) / Microcrystalline Cellulose composites were prepared by twin screw extrusion with the addition of maleic anhydride grafted polyethylene as compatibilizer and TiO_2 as pro-oxidative additives. Polyethylene wax was used as processing aid to ease the blown film process. Film was processed via conventional blown film machine. The presence of high cellulose contents had a contrary effect on the tensile properties of Cellulose-PE composite blends. However, the addition of compatibilizer to the blends enhanced the interfacial adhesion between the two materials. High content of cellulose also was found to increase the rate of biodegradability of Cellulose-PE composite films. The burst strength and soil burry test of this composite film was also increase. It suggest that this film can be used for packaging film which can degraded up to certain extend.

Keywords: Microcrystalline Cellulose, Pro-Oxidative Additives, Compatibilizer, Bio Degradable, Packaging Film

TABLE OF CONTENTS

CHAPTERS	Page No.
CHAPTER 1: INTRODUCTION	**5-13**
1.1 Introduction	5
1.2 Problem statement	10
1.3 Objectives of the study	12
1.4 Scopes of the Study	13
CHAPTER 2: LITERATURE REVIEW	**14-31**
2.1 Plastic and Environment	14
2.2 Polymer Used in Packaging	17
2.3 Development in Biodegradable Packaging materials	18
2.3.1 Cellulose Based Biodegradable Polymer	19
2.3.1.1 Cellulose Based Linear Low Density Polyethylene Biodegradable Polymers	20
2.4 Processing of Biodegradable Packaging Material	22
2.5 Properties of Biodegradable Packaging Material	22
2.6 Modification of Biodegradable Packaging Materials	23
2.7 Degradation	24
2.8 Biodegradation	25
2.9 Factors Affecting Biodegradation	26
2.10 Mechanism of Biodegradation	27
CHAPTER 3: METHODOLOGY	**32-39**
3.1 Material	32
3.1.1 Matrix	32
3.1.2 Filler	32
3.1.3 Pro-Oxidative Additive	32

3.1.4 Compatibilizer	32
3.1.5 Processing Aids	32
3.2 Experiment	33
3.2.1 Compounding	33
3.2.2 Blown Film	34
3.2.3 Compound Formulation	36
3.3 Melt Flow Index Determination	36
3.4 Mechanical Testing	36
3.4.1 Tensile Test	36
3.4.2 Water Absorption Test	37
3.4.3 Differential Scanning Calorimetry (DSC)	38
3.5 Degradation	39
3.5.1 Soil burry Test	39
CHAPTER 4: RESULTS AND DISCUSSION	**40-47**
4.1 Melt Flow Measurement	40
4.2 Mechanical Test	40
4.2.1TensileTest	40
4.3 Water Absorption Test	43
4.4 Burst strength	44
4.5 Biodegradability	45
4.5.1 Soil Burial Analysis	45
4.6 Thermal Properties	46
4.6.1 Differential Scanning Calorimetry (DSC)	46
CHAPTER 5: CONCLUSIONS	**48-49**
5.1 Final Conclusion	49
CHAPTER 6: REFERENCES	**51-54**

CHAPTER 1

INTRODUCTION

1.1 Introduction

Synthetic polymers have become technologically significant since the 1940s and packaging is one industry that has been revolutionized by oil-based polymers such as polyethylene (PE), polypropylene (PP), polystyrene (PS), poly (ethylene terephthalate) (PET) and poly (vinyl chloride) (PVC). Plastics' versatility allows it to be used in everything from the simple part, for example plastic bags, bottles and dolls to the high-tech parts, cars, computer casing, electronic devices casing and many more. The reason behind multiuse of plastics is unique capability to be manufactured to meet very specific functional needs for consumers. Plastics have been found useful in applications ranging from transportation, packaging, building, Medical appliances, agricultures and communication as shown. [1]

PLASTIC PACKAGING WASTE

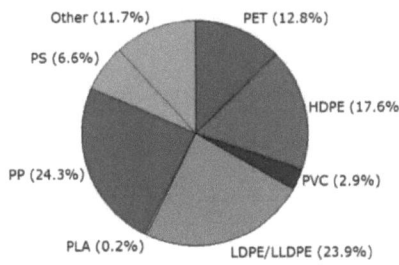

(Source: https://doi.org/10.1016/j.wasman.2015.11.041)

Worldwide production of plastics is more than 78 million tons per year and almost half of that is discarded within a short time, remaining in garbage deposits and landfills for decades (more than 30 years)[2]. Furthermore, major application of plastics is in packaging and this situation may contribute to serious environmental problems. Thermoplastics are widely used in packaging and fabrication of bottles and films [3] Synthetic plastics accumulate in nature at a rate of 25 million tons per year and polyethylene represents 64% of the produced synthetic plastics. Petrochemical based plastics such as polyolefin, polyesters and polyamides have been increasingly used as packaging materials because of their availability in large quantities at low cost and favourable functionality characteristics such as good tensile and tear strength, good barrier properties to oxygen and aroma compounds and heat seal ability [4].

However, these plastics are made of petroleum-based materials that are not readily biodegradable. Synthetic plastics such as polyethylene and polypropylene have a very low water vapour transmission rate and most importantly, so that, they are totally non-biodegradable, and therefore lead to environmental pollution, which pose serious ecological problems. Polyolefin are not degraded by microorganisms in the environment, which contributes to their long lifetime of hundreds of years. There has been an increased interest in enhancing the biodegradability of synthetic plastics by blending them with low cost natural biopolymers.

Table 1.1 Main plastics and their applications Plastics Applications

Plastics	Applications
Low density polyethylene (LDPE), linear low density polyethylene (LLDPE) polyvinylchloride (PVC)	Films and packaging
Polyethylene terephthalate (PET), PVC, high density polyethylene (HDPE)	Bottles, tubes, pipes, insulation molding
Polystyrene (PS) polypropylene (PP), PVC	Tanks, jugs, containers
LDPE, LLDPE	Bags
Polyurethane (PUR)	Coating, insulation, paints, packing

(Source: https://www.researchgate.net/publication/222370742)

Most light weight plastic packaging material is used for a one-time application and discarded when its useful life is over. These materials are durable and inert in the presence of microbes thus leading to a long term performance [5]. Although there has been a lot of new technology and method in recycling and reducing plastics waste, the number of these materials is still increasing every year. Replacing plastics to other materials such as paper and metals is less attractive because of the special characteristics and economic factors. Recycle of products also has it limitation such as high cost of operation, besides, the technology of recycling are still under development.

Many packaging materials do not lend themselves to recycling because of contamination, and the cleaning necessary prior to recycling can be very expensive. Biodegradable plastics are plastics that can undergo a degradation process known as biodegradation. They are defined as plastics with similar properties to conventional plastic but which can be decomposed after disposal to the environment by the activity of microorganism. It is also defined as plastics with similar properties to conventional plastics, but it can be decomposed after disposal to the environment by the activity of microorganisms to produce end products of CO_2 and H_2O. [4] Biodegradable plastics provide opportunities for reducing municipal solid waste through biological recycling to the ecosystem and can replace the conventional synthetic plastic products. In addition, it is desirable that these biodegradable polymers come primarily from agricultural or other renewable resources for a sustainable environment. Many synthetic materials like polyolefins are not degraded by microorganisms in the environment, which contributes to their long-life of hundreds of years [6].

Biodegradation occurs when microorganism such as bacteria and fungi degrade a polymer in an aerobic and an anaerobic environment, carbon dioxide, methane and other natural products are derived from the degradation process. Hence, biodegradation can be stated as the conversion of the constituents of a polymer to carbon dioxide/methane, microbial cellular components and miscellaneous by-products, by microorganisms [7]. Microorganisms break down the polymer chains and consume the material through several methods. Polyethylene (PE) is one of the mass produced non-degradable polymers and various types of PE are used extensively in many fields, including agricultural and packaging films. Among the polyolefins, linear low density polyethylene (LLDPE) is more susceptible to the attack of microorganisms in determined conditions. LLDPE had been a major use of plastics materials in packaging industries. Biodegradable polymers are considerably more expensive than competitive non-biodegradable polymers. New mechanisms for production and processing of synthetic polymer and natural polymer will be interesting alternatives to reduce the cost of biodegradable polymers in the market. Blending of low density polyethylene with a cheap natural biopolymer such as cellulose will enhance the biodegradability of this material. Incorporation of cellulose will accelerate the attack of microorganisms to LLDPE. Furthermore, cellulose is being a good choice since it is an abundant and low cost material in the market, so, it will reduce the cost of production of LLDPE/cellulose biodegradable polymer. Research on biodegradable plastics based on cellulose began in the 1970s and continues today at various laboratories all over the world. Cellulose satisfies the requirements of having adequate thermal stability with minimum interference of melt properties and negligible disturbance of product quality and has been considered as a material candidate in certain thermoplastic applications because it is known biodegradability, availability and low cost. The excellent physical properties of polyolefin make them suitable as packaging and film materials. Polyethylene (PE) blended with cellulose is already found to be a potential candidate to replace non-degradable thermoplastics in the areas of packaging. Cellulose is hydrophilic polymer, mainly due to the hydroxyls contains. In contrast, polyethylene is hydrophobic. Because of this totally different polar character of the polymers, they are immiscible. Addition of coupling agent will improve the incorporation of cellulose in LLDPE and also enhancing the biodegradability of the blends.

Low molecular weight plastic additives like plasticizers and fillers are usually susceptible to microbial attack. This leads to physical embrittlement of the polymer, leaving a porous and mechanically weakened the polymer. The microbes, in turn, release nonspecific oxidative enzymes that could attack synthetic polymers. Films of polyethylene/cellulose blends with compatibilizer were prepared using blow film extrusion machine. The degradation of the films under thermo oxidative treatment, ultraviolet light exposure, high temperature, high humidity and natural ambience (soil burial) were increased. Cellulose are polymers that naturally occur in a variety of botanical sources and it is a renewable resource widely available and can be obtained from different left over of harvesting and raw material industrialization. The incorporation of cellulose, as naturally biodegradable polymers with synthetic polymers, such as polyethylene will produce a biodegradable film with excellent mechanical properties, can be easily process through polymer processing techniques and biodegradable. However, due to its poor melt processability, the properties of LLDPE/cellulose blends will be affected. Plasticizers reduce the brittleness of the film by interfering with the hydrogen bonding between the lipid and hydrocolloid molecules and increase film flexibility due to their ability to reduce internal hydrogen bonding between polymer chains while increasing molecular volume [4].

1.2 Problem Statement

With the growing concern about environmental pollution, the accumulation of plastics waste needs immediate resolution. Plastics packaging has become major contribution to accumulation of plastics waste in landfills. Increasing public concern over dwindling landfill space and accumulation of surface litter has promoted the development of degradable plastics. Biodegradable plastics offer one solution to managing packaging waste. Biodegradable plastics are plastics that can undergo a degradation process known as biodegradation. Thus, in the last 20-30 years, there has been an increased interest in the production and use of fully biodegradable polymers with the main goal being replacement of non-biodegradable plastics, especially those used in packaging materials. However, although these polymers possess the required properties and can be used for the production of blown film, there are not widely used due to their high cost. Biodegradable polymers are estimated to be four to six times more expensive than polyethylene and polypropylene, which are the most widely, used plastics for packaging applications. Therefore, many research attempts have been focused on the use of natural biopolymers such as starch, cellulose, lignin and chitin, which are also fully biodegradable. In addition, these materials are also very cheap and they are produced from renewable, natural sources [8]. However, due to its poor properties, these materials are not suitable for most uses in the plastics industry. Addition of cellulose as a filler in polyethylene blend will increase the biodegradability of the film and it is suitable for packaging industry. Cellulose has become a potential use as a matrix for the development of biodegradable polymers because of its fully biodegradable properties and low cost of production. Cellulose had been widely studied by many researchers in edible films and coatings, being used to protect food products [9].

In this study, maleic anhydride grafted polyethylene will be used as processing aids to improve the strength of the LLDPE/Cellulose biofilms. High content of cellulose in the polyethylene blends will enhance the biodegradability of the LLDPE/cellulose biofilms. Cellulose is susceptible to microorganisms, thus, when these blends are deposited in the environment; various microorganisms consume the cellulose, which leaves the polymer blend in a form which is full of holes. This form enables the easier disintegration of the material into small pieces. It also increases the total surface area accessible to oxygen. As a result, the oxidation of polyethylene becomes easier. Increasing the amount of cellulose causes a decrease in both tensile strength and elongation at break. As a result, the produced materials lose their ability to produce blown films. This decrease arises from poor adhesion between cellulose and linear low density polyethylene (LLDPE) due to different polar character of cellulose and LLDPE [8]. The addition of processing aids will improve the compatibility between the two materials. Processing aids in LLDPE/cellulose blending also improve the ability of the material to be process via blow film technique. Cellulose blended polyethylene films have been reported by many researchers, but there is lack of literature on their application in food packaging.

1.3 Objectives of the Study

The main objective of this study is to develop biodegradable low density polyethylene (LDPE)/cellulose packaging film with enhanced mechanical properties via blow film extrusion process.

This objective is divided into;

(i) To determine the optimum loading of cellulose in LLDPE/cellulose blends that can give good mechanical properties for packaging and can be processed using blow film machine.

(ii) To characterize the mechanical, morphological, thermal properties of LLDPE/cellulose films before and after they are subjected to biodegradation tests.

(iii) To investigate the biodegradability of LLDPE/cellulose films.

1.4 Scopes of the Study

Scopes of this study are;

i). Compounding of LLDPE/ cellulose blends using twin screw extruder and processability studies on blow film machine. Prior to compounding, all the ingredients will be mixed using twin screw extruder. Then, the compounded samples will be blown using blow film machine to study the effect of it on the processability.

ii). Mechanical properties study of LLDPE/cellulose biodegradable films.

(a) Tensile strength and elongation at break

(b) Water absorption analysis

iii). Characterization of LLDPE/cellulose biodegradable films

(a) Melt flow index analysis after compounding process to investigate its Suitability for blown film process.

iv. Biodegradation studies of LLDPE/cellulose films

(a) Natural weathering studies

(b) Exposure to fungi environment

CHAPTER 2
LITERATURE REVIEW

2.1 Plastics and Environment

Plastics are synthetic substances produced by chemical reactions. Almost all plastics are made from petroleum. "Plastics" earned their name because they can be moulded, cast, extruded or processed into a variety of forms, including solid objects, films and filaments. These properties arise from their molecular structure. Plastics are polymers, very long chain molecules that consist of subunits (monomers) linked together by chemical bonds. Plastics, depending on their physical properties, may be classified as thermoplastic or thermosetting materials. Thermoplastic materials can be formed into desired shapes under heat and pressure and become solids on cooling. If they are subjected to the same conditions of heat and pressure, they can be remoulded. Thermosetting materials acquire infallibility under heat and pressure and cannot be remoulded. Plastics are widely used, economical materials characterized by excellent all round properties, easy moulding and manufacturing. Approximately 140 million tons of synthetic polymers are produced worldwide each year to replace more traditional materials, particularly in packaging [10].

Over 60% of post-consumer Plastics waste is produced by households and most of it as single use packaging. [3]Plastics are manufactured and designed to resist the environmental degradation and also more economical than metal, woods and glasses in term of manufacturing costs and energy required. Due to these issues, plastics resins have become one of the most popular materials used in packaging. Plastics packaging has a cycle less than a year and continuously enter the waste stream on a short turnout of time. The continuous growing of plastics industries has lead to the increase volume of plastics waste in the landfill. However, the durability, strength, low cost, water and chemicals resistance, welding properties, lesser energy and heavy chemicals requirements in manufacture, fewer atmosphere emissions and light weight are advantages of plastic materials, cause these material most preferable especially in packaging industries. Reuse strategy also has it limitation. Many plastics application are not designed to reuse because of the impurities and contamination. Food packaging, disposable diapers, medical appliances and agricultural mulch bags and covers are the most common plastics products that not suitable for reusing it.

These are examples why plastics waste could be in the waste streams very fast. Recycling of plastics after final use is possible, but plastic bags, in particular, are rarely recycled. Furthermore, the technology of sorting, collecting and recycling the plastics waste is still being developed and will cost a lot of money. Collecting and sorting used plastics is an expensive and time-consuming process. While about 27 percent of aluminium products, 45 percent of paper products and 23 percent of glass products are recycled in the United States, only about 5 percent of plastics are currently recovered and recycled. Once plastic products are thrown away, they must be collected and then separated by plastic type [11]. Most modern automated plastic sorting systems are not capable of differentiating between many different types of plastics. If plastic types are not segregated, the recycled plastic cannot achieve high remoulding performance, which results in decreased market value of the recycled plastics.

Other factors can adversely affect the quality of recycled plastics. These factors include the possible degradation of the plastic during the recycling process. Furthermore, plastics wastes that enter the waste stream were normally contaminated by dirt, food scraps and waste. Cleaning of the plastics has become one of the major problems in plastics recycling. The high volume to weight ratio of plastic means that the collection and transport of this waste is difficult and expensive also caused a problem in plastics recycling. Finally, the plastics waste will be ended up in the

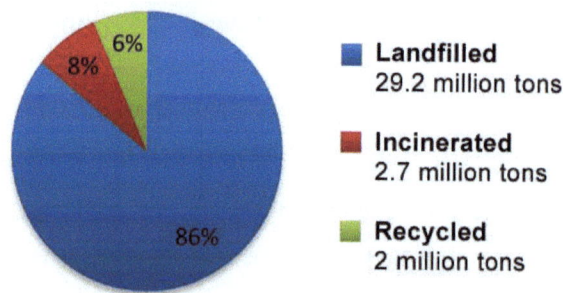

Fig 2.1 Percentage of Plastic waste recycling

(Source: http://plasticwastesolutions.com/reduce-our-plastic-usage/)

Landfill again shown in figure 2.1. Composting is the desirable method because it can be used to degrade 70% of solid waste created. However, in landfills the conditions are anaerobic and dry, not conducive to any degradation. In comparison, composting creates the optimum conditions for waste degradation; humidity, aeration and high temperatures. Plastics constitute high volume of solid waste and it is in this area that biodegradable polymers can play a major role [11].

2.2 Polymers Used in Packaging

With the growing concern about environmental pollution, the accumulation of plastic waste needs immediate resolution. Biodegradable plastics have been intensively studied in recent years and have been commercialized into various products such as garbage bags, composting yard waste bags, grocery bags and agriculture mulches [12]. Plastic packaging demand will increase more rapidly based on good opportunity for both flexible and rigid packaging. Flexible packaging advances will be fuelled by rapid growth for pouches and protective packaging. The rapidly expanding stand-up pouch segment will enable flexible packaging to gain share in a number of rigid packaging applications. In rigid plastics packaging, best opportunities are anticipated for trays, tubs and cups.

The two major applications of synthetic polymers fall in the field of food packaging (wrapping materials) and other uses, such as mulch films, seedling pots and binding twine. Plastics have gained a unique position in food packaging technology for a number of quite different reasons including; (a) higher strength, elongation and barrier properties against waterborne organisms responsible for food spoilage (b) lower cost and higher energy effectiveness (c) lightness and water resistance The continuous growth of polymer materials for food packaging applications in conjunction with their recalcitrance toward degradation and their visibility in the environment when discarded have stimulated further research in the field of food packaging [13]. It has been estimated that 2% of all plastics eventually reach the environment, thus contributing considerably to a currently acute ecological problem.

2.3 Development in Biodegradable Packaging Materials

Over the last thirty years, there has been a growing interest in biodegradable polymers. Initial interests were in the fields of medicine, such as producing degradable fibers for sutures, and agriculture, for mulch films and controlled pesticide release [11]. In more recent years attention has been focused on the rising concern for the environment. Biodegradable plastics are plastics that can undergo a degradation process known as biodegradation. [11] Biodegradation of a plastic materials leading to a change in its chemical structure caused by biological activity leading to naturally occurring metabolic end products. A plastic material is called biodegradable if all its organic components undergo a total biodegradation. Rates of biodegradation are to be determined in standardized test systems. Biodegradable plastics also defined as plastics with similar properties to conventional plastics, but it can be decomposed after disposal to the environment by the activity of microorganisms to produce end products of CO_2 and H_2O [4].

It is also an alternative to the petroleum based non-biodegradable polymers. Biodegradable plastics can be used in hygiene products, household goods, horticultural products, agriculture, medical products and many more. It decreases the solid waste problems created by plastics waste. Biodegradable polymers can be divided to two main categories, which are naturally occurring biodegradable polymers and synthetic biodegradable polymers. Naturally occurring biodegradable polymers including polysaccharides such as starch, cellulose, chitin/chitosan, pullulan, levan, konjac and elsinan. In this compound, simple sugar such as glucose, fructose and maltose are the basic units.

Synthetic biodegradable polymers are normally polymers with hydrolysable backbone or polymers that are sensitive to photo degradation. Polyester is the polymer with hydrosable backbone. Examples of polymers that in the family of polyesters are poly (glycolic acid), poly (glycolic acid-co-lactic acid), polycaprolactone, polyether-polyurethane and poly (amide-enamide). Some common synthetic biodegradable and its descriptions are listed in Table 2.1 the most attractive feature of the biopolymer-based materials is their total biodegradability. As a result they fit perfectly well in the ecosystem and save the world from growing ecological pollution caused by non-biodegradable plastics, which are essentially petroleum-based. A number of aerobic and anaerobic microorganisms have been identified for biodegradation.

2.3.1 Cellulose Based Biodegradable Polymers

Polymeric materials are generally durable and inert towards microbes, thus offering long term performance. According to the emphasis on environmental pollution problems and land shortage problem for solid waste management, such as no availability of landfills, public perception, and reduction of fertility of lands by accumulation of surface litter, environmentally degradable and 'environmentally friendly' polymers are of interest [5]. Biodegradable polymers can be further broken down into two main areas; renewable and non-renewable biodegradable polymers. Essentially renewable biodegradable polymers utilize a renewable resource, for example, a plant by-product, in development of the polymer, rather than a non-renewable, for example, petroleum based resource. There is wide variety of biodegradable polymer platforms being developed. Summarises of various biodegradable polymers, along with their generic advantages, disadvantages, potential applications shown in table 2.1[14].

Table 2.1 Synthetic biodegradable polymer

Plastic Type	Name	Description
Polyester	Polyglycolic acid (PGA)	Hydrolizable polyhydroxy acid
	Polylactic acid (PLA)	Hydrolyzable polyhydroxy acid; polymers derived from fermenting crops and dairy products; compostable
	Polycaprolactone (PCL)	Hydrolyzable; low softening and melting points; compostable; long time to degrade
	Polyhydroxybutyrate (PHB)	Hydrolyzable; produced as storage material by microorganisms; possibly degrades in aerobic and anaerobic conditions; stiff; brittle; poor solvent resistance
	Polyhydroxyvalerate (PHBV)	Hydrolyzable copolymer; processed similar to PHB; contains a substance to increase degradability; melting point; toughness; compostable; low volume and costly production
Vinyl	Polyvinyl alcohol (PVOH)	Water soluble; dissolves during composting
	Polyvinyl acetate (PVAC)	Water soluble; predecessor to PVOH; has shown no significant property loss during composting tests
	Polyethylketone (PEK)	Water soluble; derived from PVOH; possibly degrades in aerobic and anaerobic conditions

(Source: Composites: Part A 77 (2015) 1–25)

2.3.1.1 Cellulose Based Linear Low Density Polyethylene Biodegradable Polymers

Research on biodegradable plastics based on cellulose began in the 1970s and continues today at various laboratories all over the world.

Technologies have been developed for continues production of extrusion films and injection-moulded plastics containing 50% or more of cellulose. Cellulose satisfies the requirements of adequate thermal stability, minimum interference with melt properties and disturbance of product quality. Incorporation of cellulose into the synthetic polymer will increase the biodegradability of synthetic polymer when starch is consumed by microorganisms. It is believed that under a rapid enzymatic hydrolysis, cellulose will be degraded leading to a void containing matrix, reduced the mechanical properties of the plastics and might be promote the biodegradation of synthetic polymer due to the increased surface area available for interaction with microorganisms.

One of the most important polymers, both in usage and in volume of production is polyethylene. As the polymer with the greatest annual production worldwide, PE is the main component in plastic waste worldwide [8]. Polyethylene is one of common synthetic polymers of high hydrophobic level and high molecular weight [15]. In natural form, it is not biodegradable. Thus, their use in the production of disposal or packing materials causes dangers environmental problems. Many solutions have been proposed for soil waste management of plastics, like recycling, incineration, landfill disposal and degradable plastics. Recycling will not yield quality products due to heterogeneous nature of the plastics. Incineration of plastics will release toxic gases and vapours, which could prove to be a serious health hazard and use of plastic in landfill operations is least preferred because of space constraints [16]. It is increasingly felt that the best alternative would be making the plastics degradable. The importance of studying linear low density polyethylene (LLDPE) biodegradable formulations is motivating many researchers in this area.

2.4 Processing of Biodegradable Packaging Materials

There are three main classes of biodegradable polymers. The first class of material is synthetic polymers, with vulnerable groups susceptible to hydrolysis attack by microbes, such as polyesters, polyanhydrides, polyamides, polyurethanes and many more. The second class of materials is composed of naturally occurring processable bacterial polymers, such as polyhyroxybutyrate (PHB) and polyhydroxyvalerate (PHV). PHB and PHV are truly biodegradable, being attacked by a wide variety of bacteria. The third class of is blends of polymers and additives that are readily consumed by microorganisms, the classic example of this class of materials is biodegradable polyethylene and cellulose blends. The most popular method in preparation of cellulose and polyethylene blends was the conventional extrusion with the addition of processing aid to enhance the Compatibility of the two materials. LLDPE was compounded with well dried, cellulose using a twin screw extruder to obtain cellulose filled LLDPE film [6].

2.5 Properties of Biodegradable Packaging Materials

LLDPE/cellulose compounds generally present poor thermal-mechanical properties when compared with those of pure LLDPE [17]. They have studied the interfacial properties between cellulose and polyolefins to improve the hydrophilic and hydrophobic character respectively, Responsible for the poor mechanical properties. For that, the introduction of PE-g-MA copolymers in the LLDPE/cellulose formulations improved the thermal-mechanical properties of the blends. However, most of the cheap synthetic plastics have fairly good resistance to microorganisms due to low surface area, relative impermeability and high molecular weight. It is found that, various fungi and bacteria through cellulose hydrolysis study, have consumed the cellulose present on the surface of the polymer. The continues growth of polymer materials for food packaging applications in conjunction with their recalcitrance toward degradation and their visibility in the environment when discarded have stimulated further research in the field of food packaging [8]. Blends of LLDPE/cellulose have been prepared to study their properties and biodegradation rate. The current practice of disposing most plastics consists of landfills, composting and incineration. The prepared anaerobic bioreactor to stimulate those of a representative landfill. The first stage of degradation consists of partial cellulose removal and only at a later stage does slow rate degradation of LLDPE occur. Blends of LLDPE with cellulose were prepared and their mechanical properties were recorded. In general, the higher the cellulose content the worse the performance of the composite but the higher their biodegradability. A series of LLDPE cellulose have been prepared, varying in cellulose, to investigate their mechanical properties, gas/water permeability and biodegradability. The presence of high cellulose content (above 30% w/w) had an adverse effect on the mechanical properties of LLDPE/cellulose blends. Gas permeability and water vapour transmission rate increased proportionally to the cellulose content in the blend. The biodegradability rate of the blends was enhanced when the cellulose content exceeded 10% w/w.

2.6 Modification of Biodegradable Packaging Materials

The comparison of the susceptibility of pure LLDPE, LLDPE mixed with cellulose and a pro-oxidant, titanium dioxide to thermo and photo oxidation. They have found that, the LDPE-TiO_2 (with pro-oxidant) is more susceptible to thermal degradation and photo degradation than the other two materials. The pro-oxidant system enhances the thermal degradation and not the cellulose part, which might even retard the degradation. Pro-oxidant can promote photo oxidation to polyethylene chain, thus reduces the molecular weight of polyethylene chain. A reduction in molecular weight needs to take place before the material biodegrades at any appreciable rate. The oxidation is followed by cleavage of the chain [12].

2.7 Degradation

All natural phenomenon can cause materials degradation. Heat, light, short wavelength electromagnetic radiation, radioactive emissions, chemicals and interaction with bacteria, fungi can damage materials. Degradation is defined as a process that results in change in the properties of materials, which reduces the ability of the material to perform its intended function [18]. Degradation processes are categorized into several groups, such as, chemical, mechanical/physical and biological. Materials degradation such as thermal damage or chemical reactions, which are either entirely physical or chemical in nature coexist with combined forms of materials degradation such as corrosive wear [19]. Environmental conditions also exert a strong effect on materials degradation and there are three basic degradation mechanisms that can be identified which are scission of inter-monomer linkages in the backbone, scission of side chain linkages in the backbones, and ionically catalysed attack on side chains [20].

The degradation of polymers may proceed by one or more mechanisms, including biodegradation, photo degradation or thermal degradation, depending on the polymer environment and desired application. The combination of different factor from the environment such as sunlight, heat and humidity also has synergistic effects on the degradation [21]. Photo degradation of polymers is a natural phenomenon, involving mainly sunlight and oxygen, which cause deterioration of physical properties and appearance of the material. Some of the atoms within the polymer chain will absorb the energy of the photons from the sunlight and become excited, thus cause chain scission in polymer chain. Thermal effects are a major cause of deterioration of physical properties of polymers. Relatively high temperatures are often encountered in polymer processing may result in breaking chemical bonds. Polymers are sensitive to high temperature and will progressively decompose.

2.8 Biodegradation

Biodegradation is one of the several ways of polymer may degrade in the environment. This process are also interpreted by the general public as the same as other processes of polymer degradation such, as photo degradation, oxidation and hydrolysis, though they lead to very different end products. It is often conceived that the breakdown of a plastic into small, invisible fragments is biodegradation, when in reality these fragments may remain in the environment over a significant period. Biodegradable polymers when placed in bioactive environments, such as compost, will break down to carbon dioxide and water under the action of bacteria and fungi. There are two major steps in the biodegradation process. The first one involves the depolymerisation or chain cleavage of the polymer to oligomers, and the second step is the resulting mineralization of these oligomers. The depolymerisation step normally occurs outside the microorganism and involves both endo and exo-enzymes. Endo-enzymes cause random scission on the main chain, while exo-enzymes cause sequential cleavage of the terminal monomer in the polymer main chain. Once depolymerized, sufficiently small-sized oligomeric fragments are formed. These fragments are transported into the cell where they are mineralized. Mineralization is defined as the conversion of the polymers into biomass, minerals, water, CO_2, CH_4 and N_2. There are several standard test methods available to evaluate the biodegradability of plastics as listed in. Most of these test methods measure the percent conversion of the carbon from the designed biodegradable plastics to CO_2 and CH_4 (plus some CO_2) in aerobic and anaerobic environments, respectively. The absence of polymer and residue in the environment indicates complete biodegradation process, whereas incomplete biodegradation may leave polymer and/or residue as a result of polymer fragmentation or metabolism in the biodegradation process.

2.9 Factors Affecting Biodegradation

Polymeric materials were subjected to degradation by biological, chemical and/or physical actions in the environment. Generally, biodegradation involves successive chemical reactions, such as hydrolysis, oxidation with/without the aid of enzymes in living organisms. The rate of biodegradation was found to be affected by several factors [22]. Polymer's environment, organisms utilized and the nature of the polymeric materials are three main factors affecting biodegradation. All microorganisms have an optimum temperature, at which maximum growth rate occurs and thus highest enzyme kinetics exist. Discovered that an increase in the temperature of sewage in a waste water treatment plant, correlated with the increase in the rate of biodegradation of poly (hydroxylalkanoates) being tested. However, if the temperature in the environment becomes higher than the optimum temperature of a microorganism, then the denaturing of enzymes and other proteins in the microorganism takes place. In this case, the rate of biodegradation is reduced [11]. An optimum pH value also will affect the rate of biodegradation. A microorganism also needs a certain amount of nutrients from its environment to allow it to grow. Therefore, the concentration of nutrients is essential to the rate of biodegradation. Oxygen and moisture concentration also have considerable effect on rates of biodegradation in terrestrial environments [11]. One of the main problems in landfill sites is that there is lack of oxygen and moisture in the environment. If there is not enough moisture and oxygen in the environment, the microorganisms cannot growth. Nature of polymer substrate also affects the rate of biodegradation. Increased branching in polymeric materials will reduce the rate of degradation. Maximizing the linearity of the molecule reduces stearic hindrance facilitates the maximum susceptibility of the molecule to enzymatic attack and promotes microorganism assimilation [6]. Low molecular plastics are susceptible to degradation, due to the ability to transport into a microbial cell. List of factors that affecting the rate of biodegradation are shown in Fig 2.2.

2.10 Mechanism of Biodegradation

The production of biodegradable polymers is now rapidly increasing, and new biodegradable polymeric materials have been developed based on various factors, such as polymer structure, chemical/enzymatic modification, blending and mechanical treatments. Polymeric materials were subjected to degradation by biological, chemical and/or physical (mechanical) actions in the environment. Polymeric materials generally undergo these factors concurrently in the environment. Typical examples related to biodegradation are biological hydrolysis by hydrolase enzymes and oxidation by oxidoreductase enzymes. The hydrolase enzyme is responsible for the hydrolysis of ester, carbonate, amide and glycosidic linkages of the hydrolysable polymers producing the corresponding low molecular weight oligomers. The oxidoreductase enzyme is responsible for the oxidation and reduction of ethylenic, carbonate, amide, urethane, etc [19]. Hydrocarbons such as polyethylene, natural and polyisoprene rubbers, lignin and coal are first subjected to biological oxidation by oxidoreductase, such as oxygenases, hydroxylases, monooxygenases, peroxydases and oxidases in the biodegradation process. However, the degradation process precedes both by abiotic and biotic actions in the environment. Structure of the main chain polymer and the specific example of the related enzyme are shown in Table 2.3. Biodegradable polymers are generally degraded through two steps of primary degradation and ultimate biodegradation.

Primary degradation is the main chain cleavage forming low molecular weight fragments (oligomers) that can be assimilated by the microbe's hydrolysis or oxidative chain scission. Hydrolysis occurs using environmental water with the aid of an enzyme or under non-enzymatic conditions (abiotics). Oxidative scission occurs mainly by oxygen, a catalytic metal, UV light or an enzyme. Polymer chain can also be cleaved by mechanical strain such as bending, pressing or elongation. The low molecular weight fragments produced were incorporated into microbial cells for further assimilation to produce carbon dioxide and microbial cells, metabolic products under aerobic conditions. Under anaerobic conditions, methane is mainly produced in place of carbon dioxide and water [19].

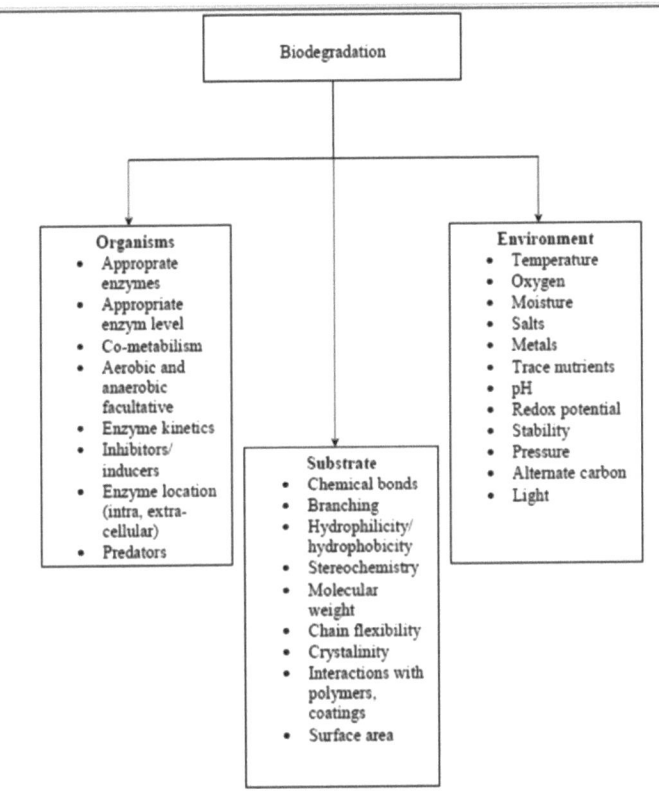

Fig 2.2 Biodegradation in plastic

(Source: Biotechnology Advances 26(3):246-65 · May 2008)

Polymer chain scission is one of the degradation phenomena in biodegradable. This process occurs in two ways, depolymerisation (exogeneous) scission and random (endogeneous) scission. In the former the polymer chain is cleaved from the terminal of the chain. A water soluble oligomer is generally liberated into the reaction media and the rate of the molecular weight reduction of the residual polymer is small. In the latter way the polymer chain is randomly cleaved. In this case, the molecular weight of the remaining polymer quickly decreased. At the same time, the mechanical properties of the remaining polymer are also quickly decreased. The addition of these two types of polymer chain scission causes degradation at the weak link. The polymer chain is cleaved at the relatively weak bond by the various physico-chemical actions. Polyesters, polyanhydrides, polycarbonates and polyamides are mainly degraded by hydrolysis into low molecular weight oligomers at the primary degradation with subsequent microbial assimilation in the biodegradation process. Some other degradation mechanisms include oxidative cleavage by a radical mechanism. Oxidative degradation is the main mechanism for non-hydrolyzable polymers, such as polyolefins, natural rubber, lignins and polyurethanes. For many polymers, hydrolysis and oxidation occur simultaneously in the environment. Surface degradation and bulk degradation are examples of polymer degradation mechanisms depending on the main degradation site. The C-C polyvinyl type polymer containing side groups, such as short alkyl Groups and phenolic groups are generally resistant to biodegradation. PVA is readily biodegradable by environmentally occurring microbes.

Polyethylene (PE) with a low molecular weight of less than 1,000 is biodegradable [23]. Biodegradation of the low molecular weight PE involves oxogenase elimination by the action of oxidoreductases, such asoxygenase, dehydrogenase and oxydase, forming a fatty acid with subsequent oxidation. The mechanism shows similarities with the typical oxidation of fatty acids and n-alkanes. For PE degradation, an initial abiotic oxidation of the polymer chain is also a necessary step. Once hydro peroxides have been introduced, a gradual increase in the ketone groups of the polymers is followed by a decrease in the ketone groups when short chain carboxylic acids are release as degradation products [23]. The combined effect of an abiotic oxidative step with consequent biotic action will be a slow but definite and progressive mineralization

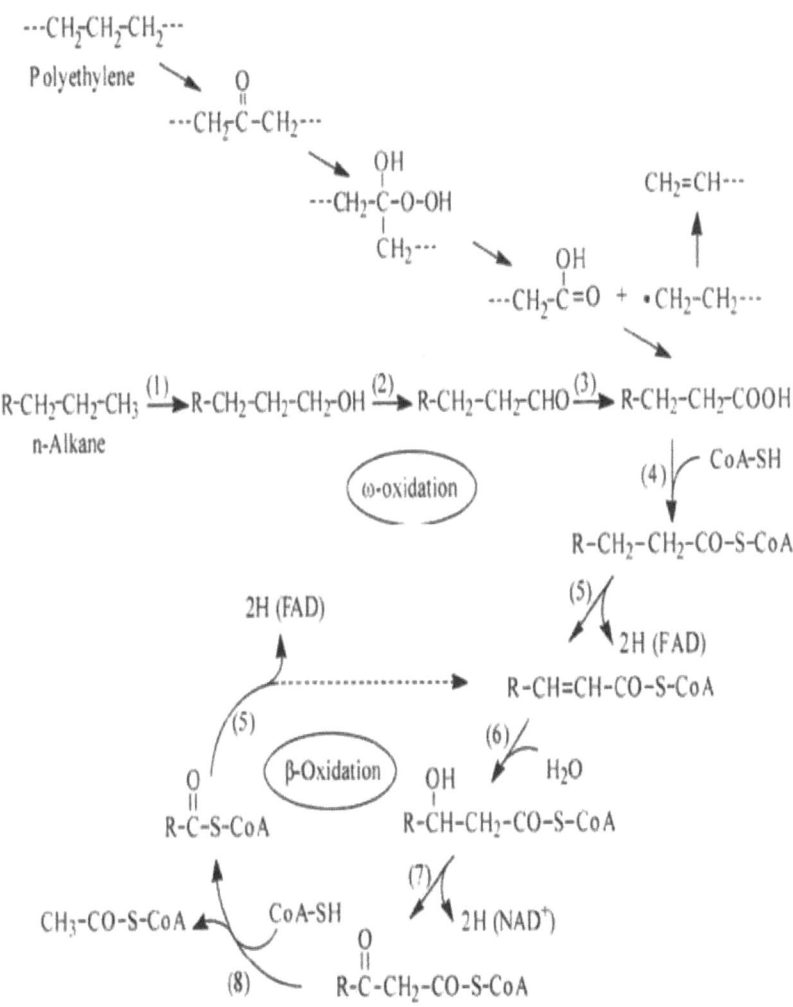

Fig 2.3 Proposed biodegradation mechanisms of polyethylene

CHAPTER 3
METHODOLOGY

3.1 Materials
3.1.1 Matrix
Low-density polyethylene (LDPE) resin grade (F19010) supplied Reliance was used in this research. The density of the polymer was 0.918 g/cm3 according to ASTM D1505. It had a melt flow of 0.90 g/10 min according to ASTM D 1238.

3.1.2 Filler
Cellulose, food grade was used as filler in this research. The cellulose used was supplied from Dariyal Polymers Pvt. Ltd. Having moisture content 10%.

3.1.3 Pro-oxidative Additive
Titanium Di Oxide was supplied by Parshwanath group of industries and the grade PA101 (Anatase) was used.

3.1.4 Compatibilizer
Maleic anhydride –grafted- poltethylene from Pluss Polymers Pvt. Ltd was used as a compatibilizer of LLDPE/cellulose blends.

3.1.5 Processing Aids
Polyethylene wax flakes (M3300) from Triveni Interchem Pvt. Ltd. Varying the amount of these processing aids to the blending will be done to get the optimum composition.

3.2 Experiment
3.2.1 Compounding

Linear Low-density polyethylene and cellulose were dried in an oven for 24 hours at 80°C before pre-mixing and compounding to dry the moisture especially for cellulose. The compounding of LLDPE/cellulose was done using twin screw extruder. The compounding process was carried out at a speed of 80 rpm and the temperature set at 140C/150C/160C/150°C. The extrudate will be palletized using a pelletizer machine for each formulation. Extrusion is a continuous process, as opposed to all other plastic production processes, which start over at the beginning of the process after each new part is removed from the mould. In the extrusion process, plastic pellets are first heated in a long barrel. A rotating screw then forces the heated plastics through a die opening of the desired shape. As the continuous plastic form emerges from die opening, it is cooled and solidified, and the continuous plastic form is then cut to the desired length. Technical specifications of twin screw extruder are listed.

Fig 3.1 Twin screw extruder

Table 3.1 Technical specifications of twin screw extruder

CIPET- AHMEDABAD	
DEPT: PLASTIC ENGINEERING	
Twin Screw Extruder and pelletizer	
MAKE	: M/s Specific Engineering & Automats
MODEL	: High speed Torque ZV 20
MACHINE SPECIFICATION	
Extruder:-	
Type	: Co Rotating
Driving Motor	: 5.5 kw
Torque/shaft	: 45Nm
Screw Dia	: 21mm
Torque	: 90 Nm
Sr.No	: SP 244CP 5
Output RPM	: 600
L/D Ratio	: 40:1
Output	: 26 kg/hr
No. of strips to be cut into dicers	:1-2
Standard granules to be cut	:3*3 mm
Cutting tool dimensions	: 90*50 mm
Cutting tool rotation speed	:0-1000 r/min

Table 3.2 Technical specifications of Blown Film Plant

3.2.2 Blown Film

The compounded samples were blown using blow film machine to produce LLDPE/cellulose plastic bags. This process will be carried out at temperature of 165^0 C/160^0C/150^0C/140^0C/130^0C/120^0C with drawer and screw speed of 50 rpm and 600 rpm respectively. Blow film extrusion is the process used to make plastic continuous sheets. This process works by extruding a hollow, sealed-end thermoplastic tube through a die opening. As flattened plastic tube emerges from die opening, air is blown inside the hollow tube to stretch and thin the tube to the desired size and wall thickness. The plastic is then air-cooled and pulled away on take up rollers.

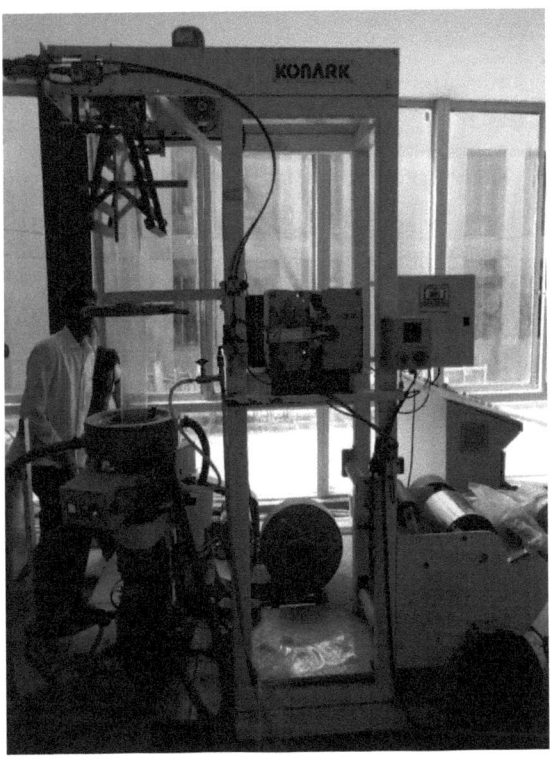

Fig 3.2 Blown film plant

3.2.3 Compound Formulation

Table 3.3 Samples formulation

MATERIALS	$C_{10}P_1$ (%)	$C_{15}P_2$ (%)	$C_{20}P_3$ (%)	$C_{15}P_0$ (%)
LLDPE	85.75	80.75	75.50	81.00
Cellulose	10	15	20	15
MAPE	3	3	3	3
PE Wax	1	1	1	1
TiO2	0.25	0.25	0.25	0

3.3 Melt Flow Index Determination

Melt flow index will be determined using Melt flow Indexer according to ASTM D1238. The temperature of 190^0C and load of 2.16 kg will be used. The time taken for the interval is one minute. The weight of extrudate will be measured and the melt flow of the samples will be calculated. The MFI corresponds to the mass of polymers that passes through a standard capillary, in an interval of 10 min, at a given applied pressure (load).

3.4 Mechanical Testing

3.4.1 Tensile Test

Fig 3.3 Tensile test Specimen

Tensile test will be carried out using an Instron machine Lloyd. The test will be done according to ASTM D 638. Gauge length will be set at 50 mm and the crosshead speed of testing will be fixed at 500 mm/min. Samples for tensile measurements will be conditioned at 30±2% relative humidity for 24 hours before testing and ten samples will be tested for each formulation. The conditioning of tensile specimens will be followed accordingly as stated by the standard. Tensile modulus, tensile strength and elongation at break will be evaluated from stress-strain data.

3.4.2 Water Absorption test

This test is carried out to study the water resistance of LDPE/cellulose films. Samples were dried at 80^0C in a vacuum oven until a constant weight was attained prior to immersion in water in a thermo stated stainless steel water bath at 30^0C. Weight gain of the samples was recorded by periodic removal of the specimens from the water bath and weighing on

$$Mt(\%) = \frac{(Ww - Wd)}{Wd} x100$$

a balance with a precision of 1 mg. The percentage of weight gain at 24hrs was taken; as a result of moisture absorption was determined by equation.

Where Wd and Ww are weight of dry material (the initial weight samples prior to exposure to the water absorption) and weight of samples after exposure to water absorption respectively. The percentage equilibrium or maximum moisture absorption Mm will be calculated as an average value of several consecutive measurements that show no appreciable additional absorption.

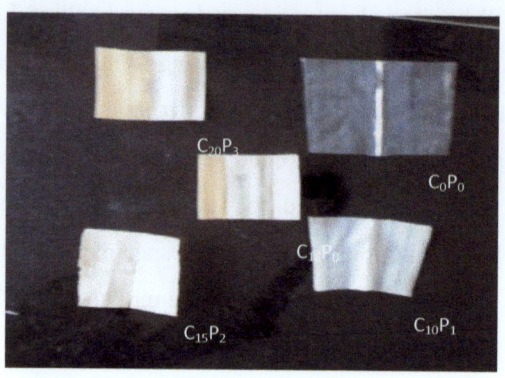

Fig 3.4 Water absorption test specimen

3.4.3 Differential Scanning Calorimetry (DSC)

DSC analysis was conducted by using a DSC. Samples of 3-5 mg were scanned from 50 to 200°C at a heating rate of 2°C/min. In the scanning, the whole samples were used for the average values of the thermal properties thus; the influence of the degraded surfaces was reduced. After being held isothermally at 200°C for 5 min, samples were cooled down at 2°C/min. Then the samples were reheated to 200°C at 2°C/min (2nd run). Crystallization temperature (Tc) and enthalpy (ΔHc) were measured from the first cooling run, while the peak melting temperature (Tm) and melting enthalpy (ΔH_m) were determined from second heating cycle.

3.5 Degradation
3.5.1 Soil burry Test

Garden soil (1200 g) was taken in different pots. A weighed amount (1 g) of each of the samples that is, PE, PE-g-Starch (composite and true graft separately) wrapped in synthetic net were placed in the beaker such that the soil covered the polymer from all the sides. The pots were covered with the aluminium foil and kept at room temperature. The weight of all the samples, PE and the grafted PE, were taken at regular interval of time (10 days) to check for any weight loss. Percent wt. loss as a function of number of days was determined.

$$\% \text{ Wt.Loss} = \frac{\text{Initial wt. at the beginning} - \text{Final wt. after ten days}}{\text{Initial wt. at the beginning}} \times 100$$

$$\% \text{ Wt.Loss (after every 10 days)} = \frac{\text{Initial wt. before ten days} - \text{Final wt. after ten days}}{\text{Initial wt. before ten days}} \times 100$$

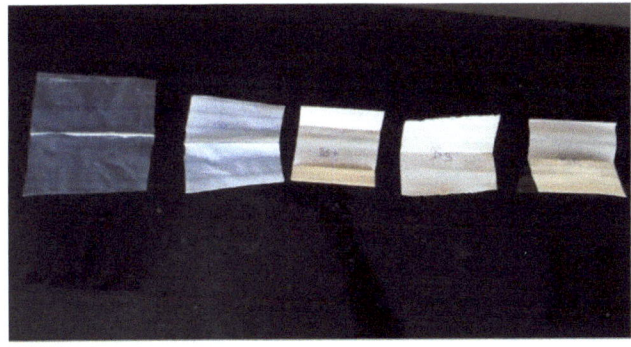

Fig 3.5 Soil burial test specimen

CHAPTER 4
RESULTS AND DISCUSSION

4.1 Melt Flow Index (MFI) Measurement

The melt flow index (MFI) values of LLDPE/Cellulose composites decreased as the content of cellulose increased. Reduction in MFI values indicate the viscosity of composite increased.

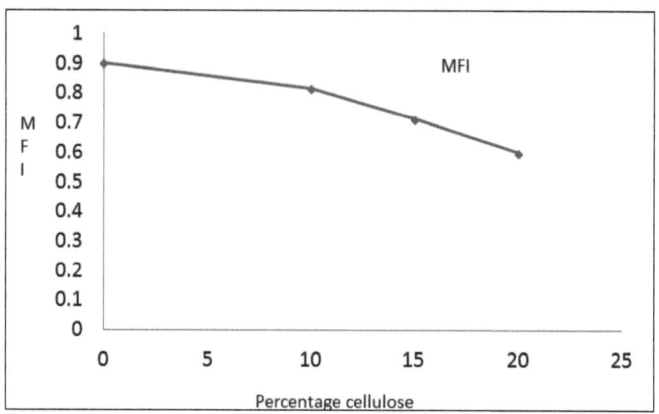

Fig 4.1 MFI values for different composition of cellulose

4.2 Mechanical Test
4.2.1 Tensile Test

Variations of tensile strength, elastic modulus and elongation at break of LLDPE/cellulose composite films had been observed. As expected, tensile strength and elongation at break of LLDPE/Cellulose blends decreased as the cellulose content increased. While the elastic modulus of the composites increased as the cellulose content rose. Similarity trend was also observed in composite films loaded with cellulose except for the tensile strength that showed an increased as the filler content increased. This was expected due to cellulose behaves as reinforcing fillers through the existence of crude fiber. Reduction of tensile strength and elongation at break was probably caused by less effective cross sectional area of LLDPE matrix (i.e. continuous phase) toward spherical particulates starch granules as the cellulose contents rose. Subsequently resulting in reduce of tensile strength.

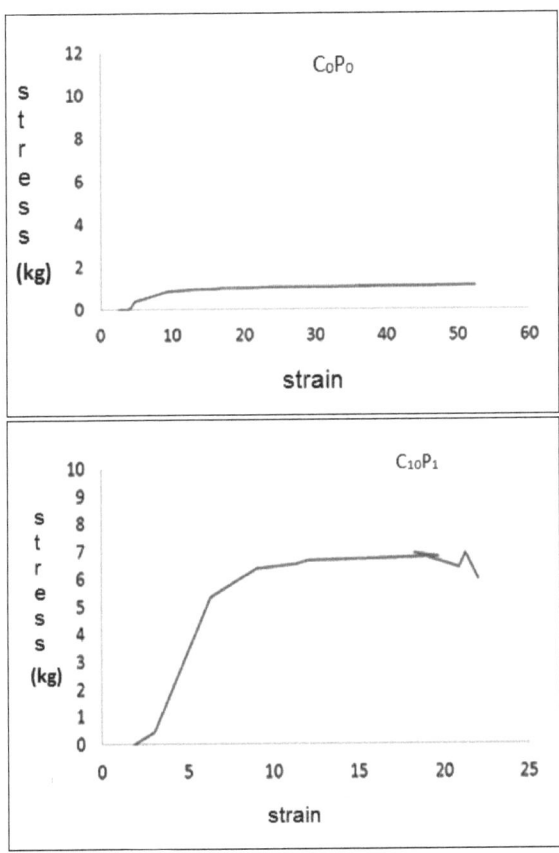

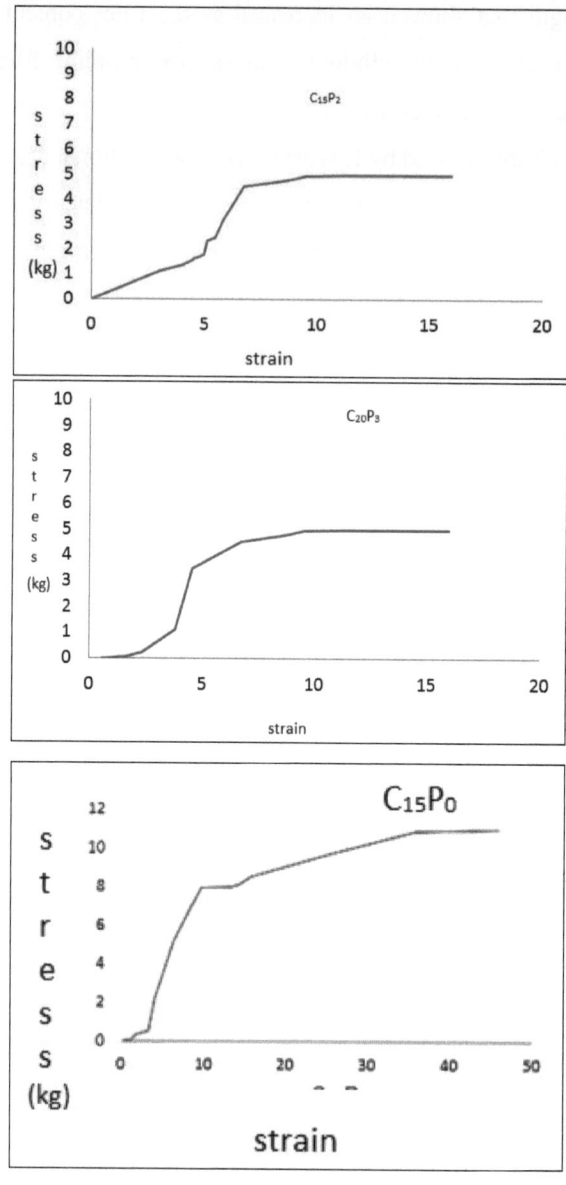

Fig 4.2 Tensile test graphs for different composition of cellulose

4.3 Water Absorption Test

For neat LDPE films, a little water uptake can be observed. Cellulose based synthetic materials tend to absorb water because the hydroxyl group in cellulose can form a hydrogen bond with water. Since the cellulose is hydrophilic, it has a highly tendency to attract water molecules. Therefore, as cellulose content increased, the tendency increases.

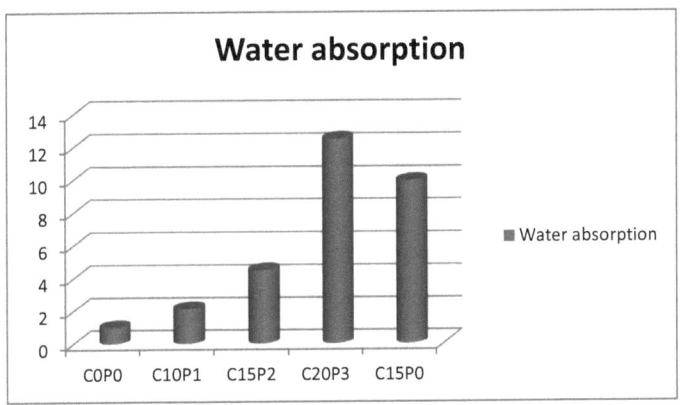

Fig 4.3 Water absorption percentage for different composition of cellulose

Composite films contain 40% and 50% of starch show high water uptake. Meanwhile, composite films containing 10%, 15% and 20% of cellulose content differ slightly from each other. This was expected due to the low concentration of cellulose particles near to the composite surfaces while the rest were positioned inter the matrix. The interior cellulose particles were not available to form hydrogen bonding with water molecules as they are trapped in LLDPE matrix. For higher loading of cellulose, cellulose particles filled and crowded the composites and resulted in higher concentration of cellulose near the composite surface. Moisture uptakes in cellulose- LLDPE composites is mainly due to the starch particles, exposed starch granules or those at or near the surface absorb moisture faster than those interior. Cellulose-PE composites take months to equilibrate even completely immersed in water. In contrast, some formulation of composite films in this study showed equilibrium achievable.

4.4 Burst Strength

The bursting strength of paper or paperboard is a composite strength property that is affected by various other properties of the sheet, principally tensile strength and stretch. Generally, bursting strength depends upon the kind, proportion, and amount of fibers present in the sheet, their method of preparation, their degree of beating and refining, upon sheet formation, and the use of additives. The test specimen, held between annular clamps, is subjected to an increasing pressure by a rubber diaphragm, which is expanded by hydraulic pressure at a controlled rate, until the test specimen ruptures. The pressure reading at the instant of rupture is recorded as the bursting strength. The units of expression are pounds per square inch or "points".

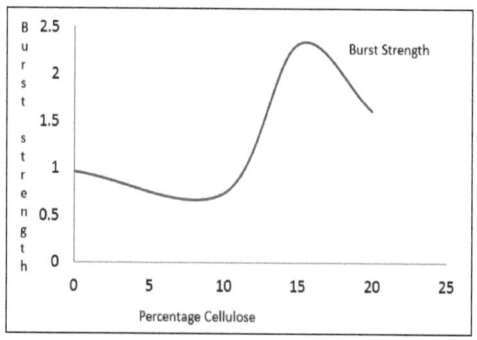

4.5 Biodegradability

4.5.1 Soil Burial Analysis

All LLDPE/cellulose specimens were exposed to moist compost for a period of 15, 30, 45 days and weight loss was recorded. All LDPE/starch Specimens were exposed to moist compost for a period of 15, 30.45 days and weight loss was recorded. The samples were not discoloured indicating that samples were not chemically interacted, but reduction in weight was observed. This is due to bleaching, dissolving or degradation of cellulose microorganism attack. The rate of biodegradation of LLDPE/cellulose blends are graphically shown in Figure. From the figure, it was noticed that the rate of biodegradation increases with increase in starch content in LLDPE matrix. The percent weight loss increased with increase in cellulose content in LLDPE system.

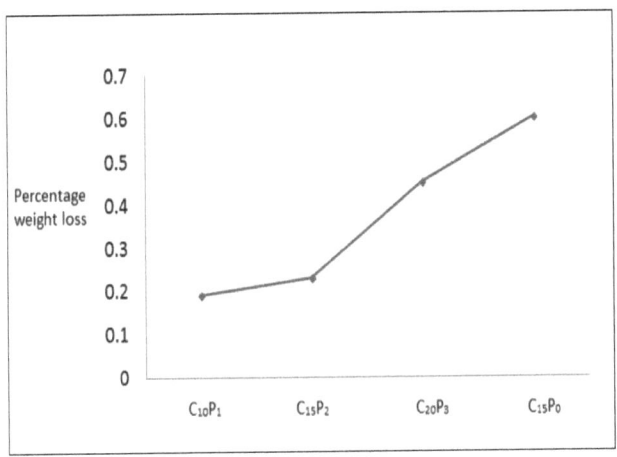

Fig 4.4 Weight loss of samples at different composition

4.6 Thermal Properties
4.6.1 Differential Scanning Calorimetry (DSC)

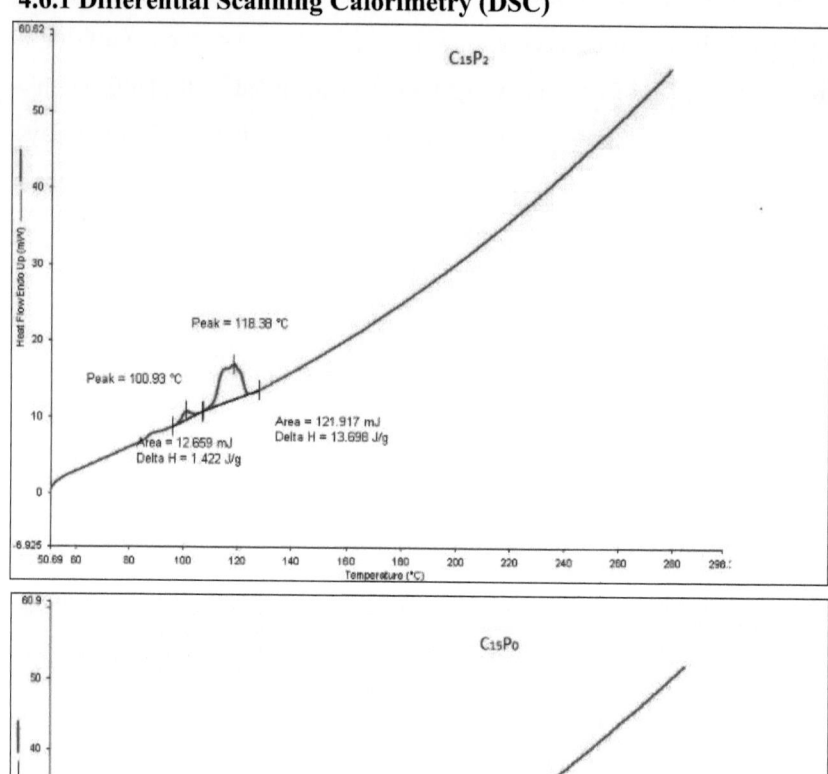

Fig 4.5 DSC curve of different composition of pro-oxidant filler A) C15P2 and B) C15P0

Melting temperature (Tm) of the composite samples were obtained from the second heating curves in the DSC scanning, two endothermic melting peak was observed for all of the samples, which was actually a characteristic two forms of crystal with infinite crystal length. The melting peak occurred (melting temperature, Tm) at $120.40°$ C and $102.82°$ C for sample without titanium di oxide while it varied from $118.38°$ C and $100.93°$C for other titanium di oxide based control composites. The presence of Pro-oxidative additive affects the thermal stability of the polymer. The thermal stability of polymer decreases with the presence of pro-oxidant additive. Pro-oxi additive decreases the thermal stability which helps in bond breaking as a much simple phenomenon than without pro-oxidative.

CHAPTER 5
CONCLUSIONS

Based on this study, there is an effective conclusion that cellulose are strongly affected the physical, chemical, and mechanical properties of LLDPE/cellulose blends. Addition cellulose has increased the biodegradability characteristics of LLDPE where in microbes assimilate the cellulose particles and leave the LLDPE matrix alone with the weaken bonding of polymer chains. Then extent to break the LLDPE chains down into small particles with a large surface area. Thus, high molar mass of LLDPE becomes lower and subsequently available to degrade consecutively.

However, the cellulose content is not effective on the mechanical properties. Cellulose imparts an adverse effect upon the mechanical properties wherein decreased the tensile strength and elongation at break, while modulus increased. This less effectiveness is due to the hydrophilic nature of cellulose that is not compatible to hydrophobic nature of synthetic polymers that result in weakness of interfacial adhesion. In contrast, incorporation of cellulose has increased the tensile strength.

The presence of filler has an effect on its burst strength as the cellulose increases burst strength increases but after a certain percentage of loading the burst strength decreases.

As cellulose loading is increased, their cohesive strength becomes stronger than particle-matrix interaction causing of high modulus and lowering of tensile strength and elongation at break. Apart of mechanical properties, this condition also contributes to reduction of melt flow index

(MFI), in such way that rigid particles restrict the flow of matrix. Cellulose has aggressive behaviour of water absorption due to the existence of hydroxyl group (OH) that is very attractive to form hydrogen bonding with water molecules.

The presence of Pro-oxidative additive affects the thermal stability of the polymer. The thermal stability of polymer decreases with the presence of pro-oxidant additive. Pro-oxi additive decreases the thermal stability which helps in bond breaking as a much simple phenomenon than without pro-oxidative.

5.1 Final Conclusion

Based on this study, there is an effective conclusion that cellulose are strongly affected the physical, chemical, and mechanical properties of LLDPE/cellulose blends. Addition cellulose has increased the biodegradability characteristics of LLDPE. Cellulose imparts an adverse effect upon the mechanical properties wherein decreased the tensile strength and elongation at break, while modulus increased. This less effectiveness is due to the hydrophilic nature of cellulose that is not compatible to hydrophobic nature of synthetic polymers that result in weakness of interfacial adhesion.

The presence of filler has an effect on its burst strength as the cellulose increases burst strength increases but after a certain percentage of loading the burst strength decreases. Apart of mechanical properties, this condition also contributes to reduction of Melt Flow Index (MFI), in such way that rigid particles restrict the flow of matrix. It found that, the LLDPE-TiO_2 (with pro-oxidant) is more susceptible to thermal degradation and photo degradation than the other two materials. The pro-oxidant system enhances the thermal degradation and not the cellulose part, which might even retard the degradation.so that this film can be used for packing material which can be degradable to some extent.

CHAPTER 6
REFERENCES

1. Steven, E. S. 2002. "Green Plastics, an Introduction to the New Science of Biodegradable Plastics." Princeton University Press: USA.
2. Volke-Sepulveda, T., Favela-Torres, E., Manzur-Guzman, Limor-Gonzalez, M. and Trejo-Quitero, G. 1999. "Microbial Degradation of Thermo-oxidized Low Density Polyethylene. Journal of Applied Polymer Science. 73. 1435-1440.
3. Zheng, Y. and Yanful, E. K. 2005. "A Review of Plastic Waste Biodegradation." Critical Review in Biotechnology. 25. 243-250.
4. Tharanathan, R.N. 2003. Biodegradable Films and Composite Coatings: Past, Present and Future. *Trends in Food Science & Technology*. 14: 71-78.
5. Albertsson, A. C, Andersson, S. O. and Karlsson, S. (1987). "The Mechanism of Biodegradation of Polyethylene." Polymer Degradation of Stabilization. 18. 73-87.
6. Arvanitoyannis, I., Biliaderis, C. G., Ogawa, H. and Kawasaki, N. 1998. "Biodegradable Films Made from Low-density Polyethylene (LDPE), Rice Starch and Potato Starch for Food Packaging Application: Part 1." Carbohydrate Polymer. 36. 89-104.
7. Chandra, R. and Rutsgi, R. (1999). "Biodegradable Polymers." Progress Polymer Science. 23. 1273-1335.
8. Raghavan, D. 1995. Characterization of Biodegradable Plastics. *Polymer-Plastic Technology Engineering*. 34(1): 41-63.

9. Bikiaris, D. and Panayiotou, C. 1998. "LDPE/starch Blends Compatibilized with PEg- MA Copolymers." Journal of applied Polymer Science. 70. 1503-1521.\

10. Flores, S., Fama, L., Rofas, A. M., Goyanes, S. and Gerschenson, L. 2007. "Physical Properties of Tapioca-starch Edible Film: Influnce of Filmmaking and Potassium Sorbate." Food Research International. 40. 257-265.

11. Brydson, J. A. (1995). "Plastics Materials" 6th Ed. Butterworth-Heinemann Ltd: London.

12. Moore, G. F. and Saunders, S. M. 1995. "Advance in Biodegradable Polymers."Rapra Review Reports. 9:2. 3-31.

13. Khabbaz, F., Albertsson, A. C. and Karlsson, S. 1998. "Trapping of Volatile Low Molecular Weight Photoproducts in Inert and Enhanced Degradable LDPE." Polymer Degradation and Stability. 61. 329-342.

14. Psomiadou, E., Arvanitoyannis, I., Biliaderis, C. G., Ogawa, H. and Kawasaki, N. 1997. "Biodegradable Films Made from Low Density Polyethylene (LDPE), Wheat Starch and Soluble Starch for Food Packaging Applications. Part 2." Carbohydrate Polymers. 33. 227-242.

15. Halley, P. J. 2005. "Thermoplastic Starch Biodegradable Polymers." In Smith, R. "Biodegradable Polymers for Industrial Applications." Woodhead Publishing Limited : England.

16. Abd el- Rahim, H. A., El-Sayid, A. H., Ali, A. M. and Rabie, A. M. 2004. "Synergistic Effect of Combining UV-sunlight-soil Burial Treatment on the Biodegradation Rate of LDPE/starch Blends." Journal of Photochemistry and Photobiology A: Chemistry. 163. 547-556.

17. Sastry, P. K., Satyaranayana, D., Rao, D. V. M. 1998. "Accelerated and Environmental Weathering Studies on Polyethylene-Starch Blend Films." Journal of Applied Polymer Science. 70. 2251-2257.
18. Nakamura, E. M., Cordi, L., Almeida, G. S. G., Duran, N. and Mei, L. H. I. 2005. "Study and Development od LDPE/starch Partially Biodegradable Compounds." Journal os Materials Processing Technology. 162-163. 236-241.
19. Charlesby, A. 1987. "Radiation Chemistry of Polymers." In Farhataziz and Rodgers, M. A. J. "Irradiation of Polymers." USA: VCH Publishers, Inc. 451-474.
20. Matsumura, S. 2005. "Mechanism of Biodegradation." In Smith, R. (Ed). "Biodegradable Polymers for Industrial Applications." Woodhead Publishing Limited:Cambridge, London.
21. Bhattacharya, M., Reis, R. L., Correlo, V and Boesel, L. 2005. "Materials Properties of Biodegradable Polymers." In Smith, R. "Biodegradable Polymers for Industrial Applications." Woodhead Publishing Limited:England.
22. Shah, P. B., Bandopadhyay, S. & Bellare, J. R. 1995. "Environmentally Degradable Starch Filled Low Density Polyethylene." Polymer Degradation Stabilization. 47. 165-173.
23. Gilmore, D. F., Antoun, S., Lenz, R. W. and Fuller, R. C. 1993. "Degradation of Polyhydroxylalkanoates and Polyolefin Blends in a Municipal Wastewater Treatment Facility." Journal of Environmental Polymer Degradation 1. 4. 269-274.
24. Albertsson, A. C. and Bahidi, Z. G. 1980. "Microbial and Oxidative Effects in Degradation of Polyethylene." Journal of Applied Polymer Science. 25. 1655-1671.

25. Albertsson, A. C and Karlsson, S. (1990). "The Influence of Biotic and Abiotic Environments on the Degradation of Polyethylene." Progress Polymer Science. 15. 177-192.